Does a Worm Have a Girlfriend?

Chicago, Illinois

Raintree

Printed and bound in the United States by Lake Book
Manufacturing, Inc.

10 09 08 07 06
10 9 8 7 6 5 4 3 2

**Library of Congress Cataloging-in-
Publication Data**
Claybourne, Anna.
  Does a worm have a girlfriend?: Reproduction /
Anna Claybourne.
     p. cm. -- (Fusion)
· Includes bibliographical references (p.   ) and
index.
  ISBN 978-1-4109-1935-9 (1-4109-1935-8) (HC)
  ISBN 978-1-4109-1966-3 (1-4109-1966-8) (Pbk)
  1. Reproduction--Juvenile literature. I. Title. II.
Series: Fusion
(Chicago, Ill.)
  QH471.C53 2005
  571.8--dc22
                                    2005011085

Acknowledgments
The author and publishers are grateful to the
following for permission to reproduce copyright
material: Alamy pp. 28–29; Corbis pp. 12–13 (Frank
Lane); Corbis/Gallo Images pp. 7, 29 (mid); Getty
Images/PhotoDisc pp. 21–22; Getty/National
Geographic p. 8; Getty/Stone pp. 26–27; Nature
Photo Library pp. 16, 29 (bottom left) (Phil Savoie);
NHPA pp. 23 (George Bernard), 15, 29 (top) (M.I.
Walker); Oxford Scientific Films p. 25 (David M.
Dennis); Photolibrary.com p. 4 (Norbert Rosing);
Science Photo Library pp. 25 (Mark Burnett), 17 (Jack
K. Clark/Agstock), 18–19, 29 (bottom right) (Andrew
J. Martinez), 5 (Ed Young/Agstock), 11 (Paul Zahl)

Cover photograph of earthworm, reproduced with
permission of Ardea/Steve Hopkin.

Illustrations by Kamae Design.

The publishers would like to thank Nancy Harris
and Harold Pratt for their assistance in the
preparation of this book.

Every effort has been made to contact copyright
holders of any material reproduced in this book.
Any omissions will be rectified in subsequent
printings if notice is given to the publishers.

The paper used to print this book comes from
sustainable resources.

Disclaimer
All the Internet addresses (URLs) given in this book
were valid at the time of going to press. However,
due to the dynamic nature of the Internet, some
addresses may have changed, or sites may have
changed or ceased to exist since publication. While
the author and publishers regret any inconvenience
this may cause readers, no responsibility for any
such changes can be accepted by either the author
or the publishers.

Contents

Some words are printed in bold, **like this**. You can find out what they mean on page 30. You can also look in the box at the bottom of the page where they first appear.

Look-Alikes

Have you wondered why babies grow up to look like their parents?

Cheetahs have baby cheetahs. Dogs have puppies. Orange **seeds** grow into orange trees. If a human has a baby, you can bet it will not be a cat, a frog, or a tomato. It will be another human. Yet why does this happen?

When cheetahs ▼ have babies, they always give birth to baby cheetahs.

4

Does a worm have a mate?

Kind of! Earthworms are different from elephants and sharks. Earthworms are both male and female at the same time!

Worms still get together to have babies. Yet each worm passes some male **sperm cells** to the other worm. Each worm joins the sperm cells with female **egg cells** in its own body. The egg cells of each worm are **fertilized**. This happens inside both worms. So, both worms are **pregnant**!

Finally, both worms lay their eggs. These eggs hatch into baby worms.

▼ These earthworms are **mating**. They swap **cells** so that male and female cells can join together. Both worms will lay eggs and have babies.

Female . . . or male?

The blackspot angelfish can start life as a female and then turn into a male halfway through its life!

Doing the Splits

Most living things use **sexual reproduction**. This is when female and male **cells** come together to form **offspring** (young). Yet there are some creatures that never need a mate. They can produce offspring on their own.

Think of an **amoeba**. It is a simple animal that has only one cell. Amoebas are tiny. They are too small to see, except through a **microscope**.

When an amoeba wants to **reproduce**, it just splits in two. Each part becomes a new amoeba. Each new amoeba can also split in two to make two more.

Amoebas reproduce on their own. They do not need a mate. This is called **asexual reproduction**.

This is a microscope picture ▶ of an amoeba. An amoeba can split in two. It can become two amoebas instead of one.

amoeba	simple animal with one body cell
asexual reproduction	offspring made by only one parent
microscope	machine that makes things look bigger

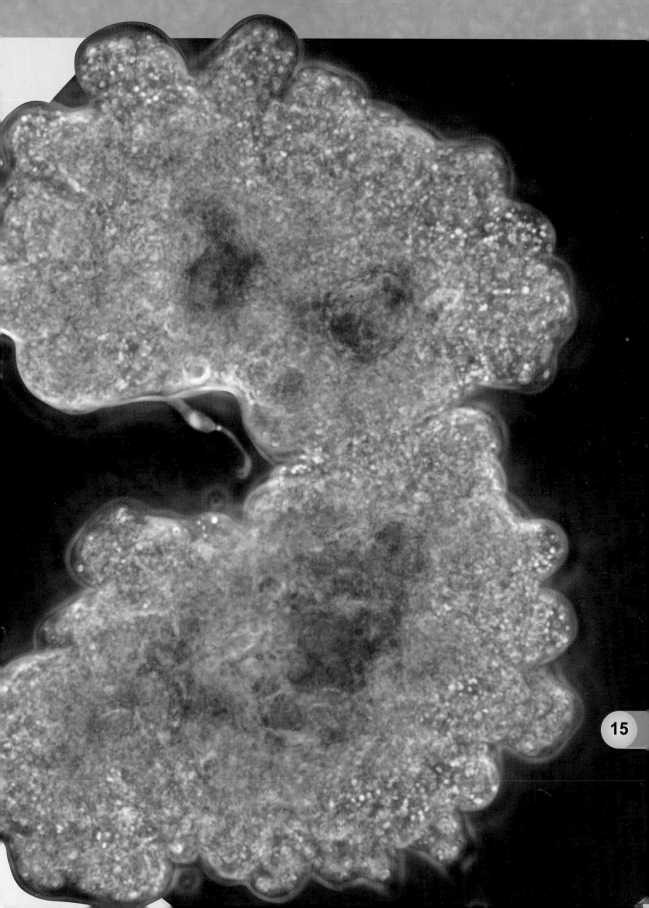

All desert ▼
grassland whiptail
lizards are females.

16

Lady lizards

Desert grassland whiptail lizards only come in one sex—female.

Each whiptail lizard lays eggs that hatch into more female whiptail lizards. This way of having **offspring** is called **parthenogenesis**. It is an unusual type of **asexual reproduction**, in which only one parent is needed to make babies. No males are needed! This is just as well, since there are no males around.

So, why can't a whiptail just **mate** with a different kind of lizard to **reproduce**? The answer is that this would not work. Mating only works with a partner of the same **species**. There are no male whiptails, so the females never mate.

Aphid alert!

Some insects, such as aphids, use parthenogenesis, too.

---

parthenogenesis   having offspring by laying eggs without mating first

A baby from an arm

If one of a starfish's arms gets broken off, guess what happens. The arm grows a new body! It grows a mouth and four more arms to make a new starfish. Meanwhile, the first starfish will grow a new arm. They are both complete starfish. One starfish has become two.

This way of **reproducing** is called **regeneration**. It is a kind of **asexual reproduction**. This means there is only one parent.

Starfish can also use **sexual reproduction**. Males and females release their **cells** into the seawater. Some of the cells bump into each other. They join together and grow into baby starfish.

Worm fact!

Flatworms are a type of worm. They can regenerate, too. You can cut a flatworm into ten pieces, and each one will grow into a new flatworm.

regeneration making offspring from a broken-off body part

▼ *This starfish is growing from an arm that has broken off another starfish.*

19

Flower Power

Plants, such as poppies, **reproduce** when male and female plant **cells** come together. They make **seeds** that can grow into new plants.

Yet how do male and female plant cells come together? Plants are rooted in the soil. They cannot move around to look for a partner.

The answer is through **pollination**. Insects land on a poppy flower to feed. They pick up male plant cells called **pollen**. When the insects visit another poppy, they leave some pollen behind. There, the pollen cells join with female plant cells. The female plant cells become **fertilized** and grow into seeds. This is pollination.

Some plants use wind to help them reproduce. The wind carries the pollen cells. Then, the cells land on other plants.

Pollen fact

*Pollen cells can only join with female cells from the same **species** (type) of plant. If pollen lands on the wrong type of flower, it will not reproduce.*

pollen male plant cells that look like yellow dust
pollination joining male pollen plant cells with female plant cells to make seeds

anther
(contains male
pollen cells)

pollen

ovary
(contains female cells)

female egg

Seed pod

Poppy flower

21

Plants use their ▲
flowers to reproduce.
After being fertilized, the
flower makes seeds.

Strawberry invasion!

If it was left alone, a strawberry plant could take over your whole garden!

Strawberry plants grow special branches, which become new plants. They are called **runners**. They creep over the soil. Then, they take root and become new strawberry plants. Finally, they separate from the parent plant.

In this way, one strawberry plant can make lots of new strawberry plants. The strawberry makes copies of itself without a mate. This is a type of **asexual reproduction**.

More strawberries!

*Strawberries can also use **sexual reproduction**. **Pollen cells** from strawberry flowers join with female strawberry cells. The female cells become **fertilized**. Then, the **seeds** form. You can see tiny strawberry seeds all over a fresh strawberry.*

22

runner   shoot that can grow into a new plant

▼ *This strawberry plant can spread out runners that turn into new plants. It can take over all the land it can reach.*

23

Runner

Passing It On

Living things pass on their own special qualities to their **offspring**. This also happens in humans.

Try this. Check your earlobes in a mirror. Do they hang down? Do they run straight across to join your head? Whatever they look like has been decided by your **genes**.

Genes are the instructions inside the **cells** of living things. They tell all living things how to grow and work.

Genes explain why offspring end up looking like their parents. The cells that make a baby contain copies of genes from both parents. So, a baby gets a mixture of instructions. Some come from its father and some come from its mother.

Because of this, everyone is the same **species** as their parents. They will even look similar to their parents. Yet each person also has his or her own special pattern of genes.

genes   instructions inside cells that make living things work

▼ *If you have "hanging" earlobes, your earlobe hangs down below where it is attached.*

◄ *If you have "attached" earlobes, your earlobe runs straight across to join your head.*

25

Your genes can make it ▲ possible for you to be naturally good at something. Yet you can still learn more and get better!

Who am I?

So, now you know how you ended up human. You had no choice! For thousands of years, humans have been having baby humans. They grew up to have more baby humans, and so on.

At the same time, **amoebas**, seahorses, sharks, lizards, and strawberries have been doing the same thing. They pass on their own **cells** and **genes**. They do this to make sure their own **species**, or type of living thing, keeps going through time.

Your surroundings and the things you do also help to make you the person you are. For example, eating healthy food makes you grow stronger. Practicing makes you better at a sport or at playing an instrument. Studying helps you learn more.

So, you can make the most of whatever you are given!

Reproduction Rundown

Type of reproduction

SEXUAL REPRODUCTION

Animal

A male and female **mate**. They join male **sperm cells** and female **eggs cells** together. This makes a baby, or an egg that can hatch into **offspring**.

Plant

A plant releases male **pollen cells**. Wind or insects carry the pollen to another plant. It joins with female cells. This makes **seeds** that can grow into new plants.

ASEXUAL REPRODUCTION

Splitting

One living thing splits into two to make two living things.

Regeneration

Part of a living thing breaks off and grows into a whole new living thing.

Parthenogenesis

A female lays eggs without mating or exchanging cells.

By runner

A plant puts out **runners** that can break off and grow into new plants.

How do they reproduce?

Can you remember which kind of reproduction these living things use to make their offspring?

A. Poppy

B. Amoeba

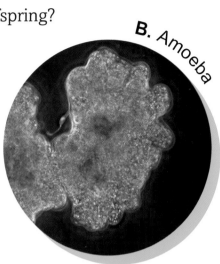

C. Elephant

D. Starfish

E. Desert grassland whiptail lizard

Answers

A. Sexual (plant)
B. Asexual (splitting)
C. Sexual (animal)
D. Sexual (animal) and asexual (regeneration)
E. Asexual (parthenogenesis)

Glossary

amoeba simple animal with one body cell. Amoebas are so small you need a microscope to see them.

asexual reproduction offspring made by only one parent. Amoebas reproduce this way. Strawberries can, too.

cells tiny units that living things are made up of. Some creatures have billions of cells, while others have only one.

egg cell female cell used to reproduce. When animals mate, egg cells join with male sperm cells.

fertilized an egg cell gets fertilized when it joins with a sperm cell

gene instruction inside cells that makes living things work. Your genes are passed on to you from your parents.

mate/mating joining body cells together to make eggs or babies. Most animals need to mate in order to reproduce.

microscope machine that makes things look bigger

offspring young that living things make when they reproduce. For example, a tiger cub is a tiger's offspring.

parthenogenesis having offspring by laying eggs without mating first

pollen male plant cells that look like yellow dust

pollination joining male pollen plant cells with female plant cells to make seeds. Pollen gets carried between plants by insects or by the wind.

pregnant used to describe an animal that has a baby or babies growing inside it. Usually, only female animals get pregnant.

regeneration making offspring from a broken-off body part. Starfish and flatworms can reproduce this way.

reproduce when a living thing makes copies of itself. All living things reproduce in order to keep their species going.

runner shoot that can grow into a new plant. Strawberry plants have runners.

seed plant part that can grow into a new plant. You can find seeds in the middle of an apple.

sexual reproduction when a male and a female cell join together to reproduce. Most animals reproduce this way.

species name for a type of living thing. Creatures always have babies that belong to the same species as themselves.

sperm cell male cell used to reproduce. When animals mate, sperm cells can join with female egg cells.

Want to Know More?

Books to read

- Spilsbury, Richard and Louise. *Life Processes: From Reproduction to Respiration*. Chicago: Heinemann Library, 2004.

- Spilsbury, Richard and Louise. *Plant Reproduction*. Chicago: Heinemann Library, 2003.

- Unwin, Mike. *The Life Cycle of Mammals*. Chicago: Heinemann Library, 2003.

Websites

- http://mbgnet.mobot.org/bioplants/main.html

  Learn more about pollination and how plants reproduce at this cool site.

To find out more about different species of living things, take a look at *Can You Tell a Skink from a Salamander?*

Did you know that different living things are specially suited to their surroundings? To find out how, read *Would You Survive?*

Index